Guide to German B-Types of WWI

Jack Herris

Great War Aviation Centennial Series #59

This book is dedicated to readers trying to identify early German biplanes.

Acknowledgements

Thanks to Mike O'Neal for the cover painting of am NFW B.I in flight.

Color aircraft profiles © Bob Pearson. Purchase his CD of WWI aircraft profiles for $50 US/Canadian, 40 €, or £30, airmail postage included, via Paypal to Bob at: **bpearson@kaien.net**

For our aviation books in print please see our website at: **www.aeronautbooks.com**.
I am looking for photographs of the less well-known German aircraft of WWI for future titles. For questions or to help with photographs please contact me at **jherris@me.com**

ISBN: 978-1-953201-03-4

Design and layout: Jack Herris
Cover design: Aaron Weaver
Digital photo editing: Jack Herris

Aeronaut Books

Books for Enthusiasts by Enthusiasts
www.aeronautbooks.com

Table of Contents

AEG B.I	3
AEG B.II	4
AEG B.III	6
AGO Pusher	7
AGO Arrow Biplane	7
AGO Tractor Biplane Trainer	8
AGO Biplane Trainer (DV 3)	8
Albatros 2-Seat Biplanes	10
Albatros *Pfeil-Doppeldecker*	14
Albatros DE	15
Albatros 1913 *Militär-Doppeldecker*	16
Albatros DD / B.I	17
Albatros B.II	18
Albatros B.III	19
Aviatik P.13 B-Type	21
Aviatik P.14 B-Type	22
Aviatik P.15 B-Type (B.I)	23
Aviatik B.II	26
Aviatik P.18 B-Type	27
Aviatik P.19 B-Type	27
Aviatik P.20 B-Type	28
DFW Mars B-Type	29
DFW B.I	30
DFW B.II	31
Euler Military Biplane Type 1914	32
Euler B.I	33
Euler B.II	34
Euler B.III	35
Fokker B.I	36
Fokker B.II	37
Germania B.I	38
Germania B.I N.D.D.	39
Goedecker B-Types	40
Gotha B.I (Gotha LD7)	41
Gotha B.II (Gotha LD2)	42
Gotha LD1	43
Gotha LD3	43
Gotha LD4	44
Gotha LD5	45
Gotha LD6	46
Halberstadt-Bristol Experimental Biplane	47
Halberstadt B-Type	48
Halberstadt B.I	49
Halberstadt B.II	50
Halberstadt B.III	51
Hansa-Brandenburg Type D	52
Hansa-Brandenburg Type FD (German B.I)	52
Hansa-Brandenburg Type LDD (Austrian B.I)	53
Jeannin Experimental B-Type	54
Jeannin Experimental 1915 B-Type	55
Knubel Experimental B-Type	56
Kondor A	57
Kondor B.I	58
Kondor W 1	59
LFG *Stahl-Pfeil-Doppeldecker*	61
L.F.G. Roland B.I	62
Lübeck-Travemünde B-type	63
LVG B.I	64
LVG B.II	65
LVG B.III	67
LVG Torpedo Bomber Prototype	68
NFW B.I	70
Otto B-Type Pusher	72
Otto B.I	73
Otto Tractor Biplane	74
Rumpler B.I	75
Sablatnig SF6 / B.I	77
SSW B.I	78
Union *Pfeil* Biplane	81
Union Racing Biplane	83
Unknown B-Type	86

Preface

While writing this series on WWI German warplanes, it was often difficult to differentiate between the multitude of unarmed German biplanes.

In Germany the early popularity of the *Taube* monoplane was overtaken by early tractor biplane designs that were lower drag and more robust. Many companies built their own biplanes to fulfill the needs of their attached flying schools, and these used similar technology and similar engines. This resulted in a large variety of unarmed biplane prototypes and a number of them in very limited production for both training and reconnaissance. At this time most B-types had side radiators; some later trainer designs were more advanced.

As air combat intensified, reconnaissance airplanes needed to be armed for self defense. Many B-types were armed, often at unit level. However, the more powerful two-seat biplanes that were designed with armament from the beginning quickly superseded the earlier B-types. Although many companies built prototype C-types (armed two-seat reconnaissance planes), fewer types were placed in production since combat effectiveness became more important and these tended to be more distinctive in appearance and thus easier to identify. At this time license production was more common for the better types.

The purpose of this book is to illustrate the identifying characteristics of German B-types to help readers distinguish between early German biplanes. Readers wanting more information, including color profiles, scale drawings, or more photos are directed to the earlier titles in the series.

Jack Herris

AEG B.I

Above: Pre-war AEG built a series of tractor biplanes powered by in-line water-cooled engines culminating in the Z6 (company designation). The German Army ordered 20 improved aircraft (Z6a) in April 1914. These were powered by a variety of 100–120 hp Benz, Mercedes, and NAG six-cylinder engines. Some served at the front from the start of the war until December 1915. In August 1915 *Idflieg* published new aircraft designations and the AEG Z6a aircraft were retroactively designated the AEG B.I. A 3-bay biplane, this one has a nose wheel to prevent nose-overs on landing. The radiator is over the engine. (The Peter M. Bowers Collection/The Museum of Flight)

Right: AEG B.I B.274, perhaps after a hard landing. The AEG B.I was a 3-bay biplane and this one had side-radiators. Application of national insignia confirms it was during the war. (The Peter M. Bowers Collection/ The Museum of Flight)

AEG B.II

Above: AEG B.II in flight. In late 1914 *Idflieg* ordered the improved AEG B.II powered by the 150 hp Benz Bz.III. This aircraft, company designation Z9, was a more compact, 2-bay development of the earlier B.I. The first production AEG B.II was rolled out in February 1915 and arrived at the front in April 1915. Some B.II biplanes were re-engined with a 120 hp Mercedes D.II and fitted with dual controls for training purposes. The diagonal cowling lines are distinctive to AEG.

Below: This view of AEG B.II B.283 shows the overall shape and many national insignia on the wings, but none on the fuselage. Side radiators were used on the B.II. (The Peter M. Bowers Collection/The Museum of Flight)

Above: AEG B.II B.244 in flight. The wing planforms are distinctive as is the rectangular horizontal stabilizer.

Below: AEG B.II B.260/14 ready for flight. Diagonal cabane struts reached to the rear of the aft cockpit. (The Peter M. Bowers Collection/The Museum of Flight)

Above: AEG B.II B.283 shows its engine and side radiator details and the ability to fold its wings for storage. The diagonal cowling border and vents are distinctive to AEG designs. (The Peter M. Bowers Collection/The Museum of Flight)

AEG B.III

Above: The AEG B.III (company designation Z10) was built in January 1915 as a lightened derivative of the B.II. Visually, the difference was a rounded fin and rudder and use of the 160 hp Mercedes D.III six-cylinder engine. As far as is known, only one B.III was built. Performance testing of the B.III compared to the B.II showed the B.III had a poorer climb despite being 10% lighter and having a slightly more powerful engine. Investigation revealed the bulged center-section gravity tank interfered with airflow from the propeller; when the tank was removed and the wing gap increased the B.III achieved the expected climb rate.

AGO Pusher

Above: The Sanke card of an Ago pusher, a license-built version of the Otto pusher. Only a few single-engine pusher designs were built in Germany and fewer were used at the front. (Reinhard Zankl)

AGO Arrow Biplane

Below: The Ago Arrow biplane (Ago no. 22) was built in 1913 and powered by a 120 hp Argus As.II. The two and a half bay wing was distinctive and finned side radiators were fitted. The cockpits were widely separated. (Peter M. Grosz Collection/STDB)

AGO Tractor Biplane Trainer

Above: The Ago Tractor Biplane Trainer was built in 1913. A 3-bay biplane, it was powered by a 100 hp Gnôme rotary. The type was also built as a floatplane. (Peter M. Grosz Collection/STDB)

AGO Biplane Trainer (DV 3)

Above & Facing Page, Top: The Ago Biplane Trainer was powered by a 100 hp Oberursel U.I rotary. It was a 2-bay biplane and the tail had no fixed fin. (Peter M. Grosz Collection/STDB)

Above: The improved Ago DV 3 was still powered by a 100 hp Oberursel U.I rotary but had a more streamlined, full cowling and propeller spinner. It had 2-bays of bracing and the tail still had no fixed fin. (Peter M. Grosz Collection/STDB)

Albatros 2-Seat Biplanes

Above: The Breguet "Tin Whistle" built under license by Albatros pre-war. The flying qualities of this design did not lend themselves to further development. (Peter M. Bowers Collection/The Museum of Flight)

Below: Another development of the Albatros MZ biplane *Taube* configuration was the MZ2 *Doppeltaube*. It retains the basic *Taube* planform on its upper wing with wing warping for roll control. (Peter M. Bowers Collection/The Museum of Flight)

Above: Another variation on the Albatros MZ biplane *Taube* configuration was the 'arrow' wing planform, characterized by sweptback wings with straight leading edges. All wings now have ailerons for roll control. (Peter M. Bowers Collection/ The Museum of Flight)

Above: During the pre-war period the *Taube* (Dove) wing planform was prized in Austria and Germany for its inherent stability, and attribute considered important to flight safety. Albatros was one of the few companies to apply the *Taube* wing to biplanes, creating the MZ1 and MZ2 in 1911. This aircraft appears to be powered by an Argus engine, making it an MZ1 *Doppeltaube*. (Peter M. Bowers Collection/The Museum of Flight)

Above: During its experimentation with the basic MZ configuration Albatros developed the tail planform; compare the horizontal tail shape above with that shown at the bottom of the facing page. (Peter M. Bowers Collection/The Museum of Flight)

Below: As Albatros continued to explore the biplane *Taube* configuration this version was built. It features conventional horizontal tail surfaces and longer-span, three-bay wings. (Peter M. Bowers Collection/The Museum of Flight)

Above: The Albatros *Pfeil-Doppeldecker* (Arrow Biplane) with 100 hp Argus. This configuration was developed from the *Taube* biplane. The inherent stability of the Taube configuration came at the expense of constrained maneuverability and high drag, which limited speed, range, ceiling, and climb rate. The arrow wing planform with ailerons on all wings was an attempt to improve performance and maneuverability. (Peter M. Bowers Collection/The Museum of Flight)

Below: This view gives a better perspective of the early arrow-wing aircraft. The horizontal tail planform was clearly derived from the earlier *Taube* configuration but uses elevators for pitch control instead of warping the tail. The wing bracing seems more than sufficient at the expense of greater drag. (Peter M. Bowers Collection/The Museum of Flight)

Albatros *Pfeil-Doppeldecker*

Above: Many early Allied designs featured rotary engines, but Germany strongly preferred the reliability and better fuel and oil consumption of water-cooled six-cylinder engines despite their weight penalty. This Albatros *Pfeil-Doppeldecker* appears to use an Argus engine. Large side radiators are fitted, creating drag and indicating the relative lack of efficiency compared to later radiators. (Peter M. Bowers Collection/The Museum of Flight)

Above: Lineup of pre-war Albatros-built aircraft. Two early DE biplanes are in front. The third aircraft in line is a license-built Breguet; an F2 based on the Farman is at the far right. (Peter M. Bowers Collection/The Museum of Flight)

Albatros DE

Above & Below: The Albatros DE displaying its wood veneer fuselage and nose radiator. The lower wings are suspended below the fuselage.

Albatros 1913 *Militär-Doppeldecker*

Above: The next step in development of the new Albatros biplane was the 1913 Albatros *Militär-Doppeldecker* (Military Biplane). The nose radiator has been replaced by a radiator over the engine and the vertical tail has been redesigned and looks more robust. The cabane struts have been simplified and the lower wings are now attached directly to the fuselage, eliminating a number of drag-producing struts. The wood veneer fuselage has been retained. The wing cellule had 2½ bays of bracing. (Peter M. Bowers Collection/The Museum of Flight)

Left & Above: The 1913 Albatros *Militär-Doppeldecker* (Military Biplane) used a 'brow' radiator and the wings were attached to the lower fuselage longerons. The wing cellule had 2½ bays of bracing.

Albatros DD / B.I

Above: This Albatros DD was photographed in December 1913 at Johannisthal, location of the main Albatros factory. This example was powered by a 100 hp Mercedes and had a Windhoff radiator above the engine. However, a variety of engines were used.Its 3-bay strut bracing was much simpler than earlier Albatros biplanes. The Albatros DD was later designated the Albatros B.I. The aircraft above has the early, low profile fin and rudder. A taller fin and rudder were applied to later B.I production aircraft. (Peter M. Bowers Collection/The Museum of Flight)

Above: Albatros DD, later designated B.I, with 100 hp Mercedes with brow radiator and the early low-profile fin and rudder.

Above: Albatros B.I B.820/14 had the later, taller fin and rudder.

Above: Albatros B.I S.69 was converted to conduct torpedo-dropping trials and is seen here with the small-caliber torpedo it was designed to carry. The landing gear was modified to enable the torpedo to be carried. The trials were unsuccessful and all later WWI German torpedo bombers were floatplanes. (Peter M. Grosz Collection/SDTB)

Albatros B.II

Above & Facing Page, Below: The Albatros B.II was essentially the same as the B.I but had a 2-bay wing. The reduced weight and drag of its smaller wing gave the B.II better performance than the B.I. Like the B.I, the B.II used a variety of engines. Early production B.II aircraft had the low profile vertical fin and rudder like early production B.I aircraft. The B.II on the facing page was powered by a Benz engine. Above, *Goslar* is in pristine condition despite serving in a frontline unit.

Albatros B.III

Above: Albatros B.III without serial number markings, probably at the factory. Side radiators were used. The most distinctive feature of the 2-bay B.III was its revised tail surfaces. (Peter M. Grosz Collection/SDTB)

Above: The Albatros B.III served in modest numbers; its primary contribution was to the design of the successful and widely-produced Albatros C.III. The Albatros logo is on the upper fin near the rudder.

Below: Albatros B.III with modified, balanced elevators. (Peter M. Grosz Collection/SDTB)

Aviatik P.13 B-Type

Above: Pre-war Aviatik B-type with uncovered fuselage. The 4-wheel landing gear had skids to prevent nose-overs; these were omitted from later Aviatik designs.

Right: Pre-war Aviatik P13 B-type was the first of a series of successful Aviatik two-seater designs. The wings were swept back and had 2½ bays of bracing.

Right: Pre-war Aviatik P13 B-type in flight. The landing gear skids to prevent nose-overs were omitted from later Aviatik designs.

Aviatik P.14 B-Type

Above: Typical Aviatik P14 B-type with simplified landing gear. The wings were swept back and had 2½ bays of bracing like the P13. Like nearly all B-types the pilot sat in the back seat to give the observer the best view forward and downward. The engine was a 100 hp Argus As.I.

Above: A pristine Aviatik P.14 with no national insignia, probably pre-war. (Peter M. Grosz collection, STDB)

Aviatik P.15 B-Type (B.I)

Above & Below: Aviatik P15 B.558/15 represents the final configuration of the Aviatik B-type P15 model with 120 hp Mercedes D.II engine. It was captured by the French in September 1915 in excellent condition and photographed for posterity. B.558/15 wears its factory markings and finish with no unit or personal markings. The P15 was a more advanced design than the P14 with its 2-bay wing without sweep-back and fixed vertical fin. It was the Aviatik B.I.

Above: Armed Aviatik B.1326/15 *Kismet* of *FFA* 6b ready for its next mission. (Bruno Schmäling)

Left: Well-used Aviatik P.15 B.192a/13 with national insignia and in unit colors. (Peter M. Grosz collection, STDB)

Facing Page: Armed Aviatik B of *FFA* 8b and its crew. (Bruno Schmäling)

Left: Aviatik P.15 B.187/14 with unit colors and retrofitted with armament. (Peter M. Grosz collection, STDB)

Reschke

Aviatik B.II

Above & Left: Aviatik B-types in the production batch 1320–1355/15 were definitely designated as Aviatik B.II. These aircraft had 120 hp Mercedes D.II engines and, with their greater payload, were used for bombing. B.II 1329/15 is shown above and 1323/15 is at left.

The Aviatik P15 B-type appears to have the final configuration of the Aviatik B and be retroactively designated the Aviatik B.I. These aircraft had a variety of engines in the 100 hp category. The aircraft definitely designated as the Aviatik B.II had 120 hp Mercedes D.II engines in the same airframe.

Aviatik P.18 B-Type

Above: The Aviatik P.18 prototype had two bays of bracing. It was fitted with a 150 hp Benz Bz.III; another version had an Argus As.II engine. One P.18 was converted to a floatplane. (Peter M. Grosz collection, STDB)

Aviatik P.19 B-Type

Above: The 2-bay Aviatik P.19 was a competition aircraft powered by a 130 hp Salmson M 9 water-cooled radial engine. The airframe was apparently derived from the Aviatik P.15. French pilot Gilbert flew it in the Star flight to Monaco in 1914, but he had to land at Lyon because of engine problems. Note the large fin. Peter M. Grosz collection, STDB)

Aviatik P.20 B-Type

Above & Below: The Aviatik P.20 racing plane with 100 hp Gnôme rotary engine. The airframe design was based on the P.15 design but was much smaller with only 10.50 meter span and had a 2-bay swept wing. The aircraft below had a revised engine cowling and curved rudder without fixed fin. It participated in the Prince Heinrich Flight competition in 1914. (Peter M. Grosz collection, STDB)

DFW Mars B-Type

Above: Ground crew attempt to steady an early DFW Mars biplane before takeoff. The engine has four cylinders. The Mars biplane was a pre-war trainer used at the front in the early months of the war. (Peter M. Bowers Collection/Museum of Flight)

Below: The DFW Mars biplanes were powered by a variety of four and six-cylinder engines such as the 95 hp NAG, etc. This later aircraft has a six-cylinder engine. The wings were swept and had 3½ bays of bracing and the landing gear had 4 wheels.

DFW B.I

Above: This DFW B.I carries national insignia on both sides of all wings. (Peter M. Bowers Collection/Museum of Flight)

Right: This DFW B.I shows its distinctive curved wing planform. Bracing was 3 bays.

Below: Ground crewmen steady the wings of a DFW B.I before take off. Like most B-types the DFW B.I had side radiators. (Peter M. Grosz Collection/SDTB)

DFW B.II

Above: DFW B.II B.641/14 served as a trainer. The DFW B.II had a 2-bay wing with rectangular planform and distinctively shaped fin and rudder. There was little resemblance to the DFW B.I other than the side radiators. (Peter M. Grosz Collection/SDTB)

Above: DFW B.II B.655/14. (Peter M. Grosz Collection/SDTB)

Euler Military Biplane Type 1914

Above: Euler military biplane type 1914 with nose wheel. The Euler had a distinctive curved engine cowling that resulted from the metal frame of the forward fuselage. This was the predecessor of the Euler B.I.

Below: Euler military biplane type 1914 in front of the Euler Werke. Note the nose wheel. (P. M. Grosz Collection/STBD)

Euler B.I

Above: Euler B.I, B43/14. The Euler B.I was a license-built LVG B.I as modified by Euler and had a distinct curved cowling. The radiator was behind and below the 100 hp Mercedes engine. The aileron 'kink' typical of early LVG designs was eliminated.

Right: Euler B.I trainer with crew.

Below: Euler B.I trainer in the snow. (P. M. Grosz Collection/STBD)

Euler B.II

Above & Below: Euler B II trainers. The B.II was later production series of the B.I with essentially the same engine and airframe. The designation may have been given to bridge the logical gap between B.I and B.III. (P. M. Grosz Collection/STBD)

Euler B.III

Above & Below: Euler B III after the correction of the tail heaviness. The Euler B.III was based on the LVG B.III with some detail changes by Euler to correct the tail heaviness that Euler found was a problem with the original LVG design. Production LVG B.III trainers were modified per Euler's findings. The engine used was a 100 hp Mercedes.

Fokker B.I

Above: Fokker B.I with a black/white fuselage band in the field surrounded by aircrew. The B.I was the military designation for the Fokker M.10E single-bay biplane. The engine was an 80 hp Oberursel U.0. (Peter M. Grosz collection/STDB)

Below: Fokker M.10E fuselage fitted with 100 hp Mercedes fitted because the M.10E was underpowered with the 80 hp Oberursel. (Peter M. Grosz collection/STDB)

Fokker B.II

Above: In service the 2-bay Fokker M.10Z was known as the Fokker B.II. The Fokker B.I and B.II shared the same fuselage. This is w/n 199. (Peter M. Grosz collection/STDB)

Above: Fokker B.II (M.10Z) 2-bay version had more wing area than the B.I for better performance. A few like this example retrained the 80 hp Oberursel U.0 with 7 cylinders; most had the 9-cylinder Oberursel U.I with 100 hp. (Peter M. Grosz collection/STDB)

Germania B.I

Above: Germania B.I front view. This aircraft has a long, over the wing exhaust. The 120 hp Argus As.II was the engine normally used for the Germania B.I. Side radiators were used.

Below: Rear view of the Germania B.I shows its conventional 2-bay configuration. (Peter M Grosz collection/STDB)

Germania B.I N.D.D.

Above & Below: This Germania B.I has an enlarged fin. The writing "1. Flug." on the photo below indicates one aircraft, which may mean one aircraft was so modified. This modification may have been a test-bed for the more powerful armed KDD type that needed a larger fin and rudder for directional control. The KDD fin and rudder were larger than the one on this aircraft due to the more powerful 150 hp Benz Bz.III used in the KDD. (Peter M Grosz collection/STDB)

Goedecker B-Types

Above: Experimental Goedecker B-type. This first version was a single-bay biplane. It was powered by a 120 hp Mercedes D.II engine cooled by side radiators. Apparently it remained a single prototype. (Peter M. Grosz collection, STDB)

Above: The experimental Goedecker B-type was rebuilt as a two-bay biplane apparently with slightly longer fuselage. Power remained a 120 hp Mercedes D.II engine. Side radiators were used. (Peter M. Grosz collection, STDB)

Gotha B.I (Gotha LD7)

Above & Below: The Gotha B.I was a conventional, two-bay biplane powered by a 120 hp Mercedes D.II engine. The fin was a distinctive shape. As usual for B-types, side radiators were used. Of the 18 aircraft built, 13 were sent to *FEA* 3 at Gotha for use as trainers and the other 5 were sent to the German Air Service in Turkey. The factory designation was LD7.

Gotha B.II (Gotha LD2)

Above: The prototype Gotha LD2 showing its streamlined engine cowling and spinner and very low profile fin and rudder.

Above: The production prototype of the Gotha LD2 (military B.II) was far less appealing than the first prototype. The cowling was modified for better cooling and the fin and rudder were enlarged for better stability and control. Nine production aircraft were built. A 100 hp Oberursel U.I provided the power.

Gotha LD1

Above: The Gotha LD1 was the start of Gotha's series of single-engine landplane biplanes. Powered by a 100 hp Mercedes D.I engine, it had a reasonable performance for its time but only one was built. Used at Gotha for training, it was commandeered by officers of *Feld-Flieger-Abteilung* 4 on 31 August 1914 and subsequently purchased by the *Fliegertruppe*, which assigned it the designation B.458/14. Only one LD1 was built.

Gotha LD3

Below: The Gotha LD3 was a license-built Caudron; only one was built. The engine was a 50 hp Gnôme.

Gotha LD4

Above & Below: The Gotha LD4, a license-built Caudron, had a spinner and streamlined nose similar to that of the prototype LD2. The engine was a 100 hp Gnôme rotary. It has side-by-side seating. Ordered in February 1914, only one aircraft was built.

Gotha LD5

Above & Below: The Gotha LD5, powered by a 100 hp Oberursel U.I, was designed as a 'cavalry biplane', a light observation plane designed to accompany an army on the move. First flown in December 1914, it was a small, 2-bay biplane with no fixed tail surfaces. The sole LD5 built served as a trainer at *FEA* 3.

Gotha LD6

Above & Below: The ill-fated original LD6 with small fin and rudder. Powered by a 150 hp Benz Bz.III engine cooled by side radiators, the 3-bay LD6 was a larger, more powerful aircraft than its predecessors. It crashed on its maiden flight on 30 January 1915, killing test pilot Oswald Kahnt. The modified Gotha LD6a with larger tail surfaces and other improvements was ordered in February 1915. Failing to meet the acceptance specifications, the LD6a was assigned to the *Herzog Carl Eduard Fliegerschule*, where it received school number 70.

Halberstadt-Bristol Experimental Biplane

Above & Below: Halberstadt-Bristol experimental biplane being manhandled by its ground crew. The engine was a 100 hp Oberursel U.I 9-cylinder rotary. This 3-bay biplane had a circular fuselage made of wood. One example was built in the beginning of 1914. (Peter M. Grosz collection/STDB)

Halberstadt B-Type

Above: This is the Halberstadt B-type work number 117, the original *Hindenburg* (or *Hindenburg I* after the second aircraft was built). The work number is barely visible on the rudder. The engine was an 80 hp Uberursel U.O seven-cylinder rotary. The wings had 2½ bays of bracing.

Above: This is the Halberstadt B-type work number 118, the *Hindenburg II*. The movable rudder has been reduced in size and is a different shape from the original *Hindenburg*. The work number is barely visible on the rudder.

Halberstadt B.I

Above: The Halberstadt B.I was significantly move advanced than the Halberstadt A.I monoplane. It was a two-seat , 2-bay trainer and reconnaissance biplane powered by an 80 hp Oberursel U.0 seven-cylinder rotary engine. The fuselage was based on the rectangular, fabric-covered steel tube fuselage from the Fokker M.8 to which was added a biplane wing cellule. (Peter M. Grosz collection/STDB)

Below: Halberstadt B.I fitted with a flexible machine gun. Other experiments fitted the aircraft with a fixed, synchronized machine gun for the pilot.

Halberstadt B.II

Above: The Halberstadt B.II had a 100 hp Mercedes D.I with side radiators. (below: Peter M. Grosz collection/STDB)

Above: The Halberstadt B.II was a trainer and reconnaissance plane. It was a 2-bay biplane with no fixed tail surfaces. Later it was developed into the Halberstadt fighter series. (Peter M. Grosz collection/STDB)

Halberstadt B.III

Above: Halberstadt B.III used for reconnaissance at the front had a larger 2-bay wing than the B.II. It was powered by a 120 hp Mercedes D.II six-cylinder engine with side radiators. In fact, it was essentially a B.II adopted for a more powerful engine; like the B.II it had no fixed tail surfaces. (Peter M. Grosz collection/STDB)

Above: Pristine Halberstadt B.III at a training facility. (Peter M. Grosz collection/STDB)

Hansa-Brandenburg Type D

Above: The Brandenburg Type D had 2-bays of bracing and side radiators. Influenced by Albatros designs, the 12 aircraft built for the German Army in 1914 were powered by 100/110 hp Benz engines

Hansa-Brandenburg Type FD (German B.I)

Above: The inward-sloping struts are a Brandenburg characteristic. Nine FD biplanes were reported to have been supplied to the German Army in 1914. Engines were quoted at 100/110 hp Argus, Benz, and Mercedes. The horizontal tailplane is in two halves and not resting on the top of the fuselage turtleback as in later Brandenburg types. Brandenburg B.I series 76, 176, 276 were from Fischamend aircraft production; at the rear the typical water tower, erected during WWI, still exists today and is in commercial use.

Hansa-Brandenburg Type LDD (Austro-Hungarian B.I)

Above: Brandenburg B.I 05.36 with typical inward-sloping struts. The aircraft was powered by a 100 hp Mercedes engine with under-wing radiator. A gun ring is mounted high over the rear cockpit of this machine.

Above: Brandenburg B.I 05.26 on a snow-covered airfield. From 05.26 the pilot was relocated from the back to the front cockpit. The B.I was visually similar to the later, more powerful and well-known Brandenburg C.I.

Jeannin Experimental B-Type

Above: Jeannin experimental 2-bay B-type. The inline engine is cooled by side radiators. This simple design appears to be the first biplane designed by Jeannin. The aircraft was not accepted by *Idflieg* and no military designation was assigned. (Peter M. Grosz collection, STDB)

Jeannin Experimental 1915 B-Type

Above & Below: This Jeannin experimental B-type appears to be the second Jeannin biplane design. The complex wing bracing is sturdy, skids are attached beneath the wing tips, and national insignia appear on both sides of the wings on white backgrounds, plus the rudder, but not on the fuselage. The engine appears to be a 150 hp Benz Bz.III. The aircraft was not accepted by *Idflieg* and no military designation was assigned. (Peter M. Grosz collection, STDB)

Knubel Experimental B-Type

Right: After the Knubel *Taube*, Knubel also built a B-type biplane with ear radiators. The 2-bay bracing was augmented by extra wire bracing. (Peter M. Grosz collection, STDB)

Below: This 1915 Knubel biplane was also given Cellon covered wings but the fuselage was fabric covered. It is thought that Knubel lost his life in this machine on 8 September 1915. (Peter M. Grosz collection, STDB)

Kondor Type A Biplane

Above & Below: Kondor A. The two-bay Kondor A was designed as a trainer and was powered by a 120 hp Mercedes D.II cooled by side radiators. Empty weight was 806 kg and gross weight was 1150 kg. The Kondor A was built in small numbers for use at the Kondor flight school. The aircraft was not accepted by *Idflieg* and no military designation was assigned. (Peter M. Grosz collection, STDB)

Kondor B.I

Above & Below: Kondor B.I two-bay trainer. The images show the side radiator was on the starboard side only. The aircraft is liberally decorated by the letter 'K' in a circle. As far as is known, only one prototype was built. (Peter M. Grosz collection, STDB)

Kondor W 1

Above & Below: Kondor W I 2-bay trainer built for the Kondor flying school. The radiator is under the upper wing center section. The Kondor W 1 trainer below is in front of the Kondor factory. The fuselage was covered with plywood and the nation insignia was painted on both sides of the upper wing but only the bottom of the lower wing The wing bracing is unusual. (Peter M. Grosz collection, STDB)

Above & Below: Kondor W 1 trainer. The 2-bay wing bracing was unusual. The engine was a 120 hp Mercedes D.II. Span was 13 m, length 7.95 m, gross weight 900 kg. A ceiling of 4,000 m was demonstrated. (Peter M. Grosz collection, STDB)

LFG *Stahl-Pfeil-Doppeldecker*

Above: LFG *Stahl-Pfeil-Doppeldecker*. The 2½-bay wing bracing was unusual. There was no fixed fin. Side radiators were fitted to cool its six-cylinder engine.

Above: LFG *Stahl-Pfeil-Doppeldecker*. The 2½-bay wing bracing was unusual. With its fixed fin this appears to be a later variation on the design. The pilot's control wheel is visible in the rear cockpit.

Roland B.I

Above: The Roland B.I was the Albatros B.II built under license by Roland. It was later designated Alb. B.II(Rol). Here is one of *Feldflieger Abteilung* 21 being inspected by *General* von Emmich in the Spring of 1915. Unarmed from the factory, a machine gun has been added at the unit for use by the observer in the front seat.

Above: Roland B.I, later designated Alb. B.II(Rol). An abundance of national insignia were applied to prevent troops shooting at their own aircraft. Twelve of these aircraft (870–881/14) were built by Roland.

Lübeck-Travemünde B-Type

Above: Lübeck-Travemünde B-type was designed as a trainer. The rudder and fin design resemble the F4 and final production batch F2 vertical tail surfaces. A wing leading-edge radiator was fitted and it had 1918 insignia. It remained a prototype. (Peter M. Grosz collection/STDB)

Above: Not a B-type but a decoy designed to look like an *Eindecker*. Note the 1918 insignia.

LVG B.I

Above: The LVG *Doppeldecker* System Schneider trainer shown here was designed by Franz Schneider. It met the Army's requirements for a two-seat reconnaissance aircraft and was very successful. The orders for this aircraft established LVG as one of the most important German aircraft manufactures. This 2-bay aircraft, retroactively designated the LVG B.I, became the foundation of LVG success and was the most numerous of the aircraft types with which the *Fliegertruppe* started the war. This one was powered by a 110 hp Benz Bz.II and was fitted with side radiators like most B-types.

Below: The early LVG B.I below had a nose wheel and skid to prevent nose-overs and landing. These features were soon eliminated in favor of the simplified landing gear with reduced weight and drag. The wings were slightly swept.

LVG B.II

Above: LVG B.II B.723. The LVG B.II was a slightly smaller, more compact aircraft than the B.I that offered improved performance and maneuverability. The B.II had a cutout in the rear of the upper wing for better field of view for the pilot; the B.I did not have a cutout and that is the key way to distinguish between the two types at are otherwise nearly identical. (Peter M. Grosz Collection/STDB)

Below: LVG B.II(Ot) 1510/15 is an Otto-built machine at the front. Eight of this type (1503/15–1510/15) were built. It had rails on both sides of the observer's cockpit for mounting a machine gun. (Peter M. Grosz Collection/STDB)

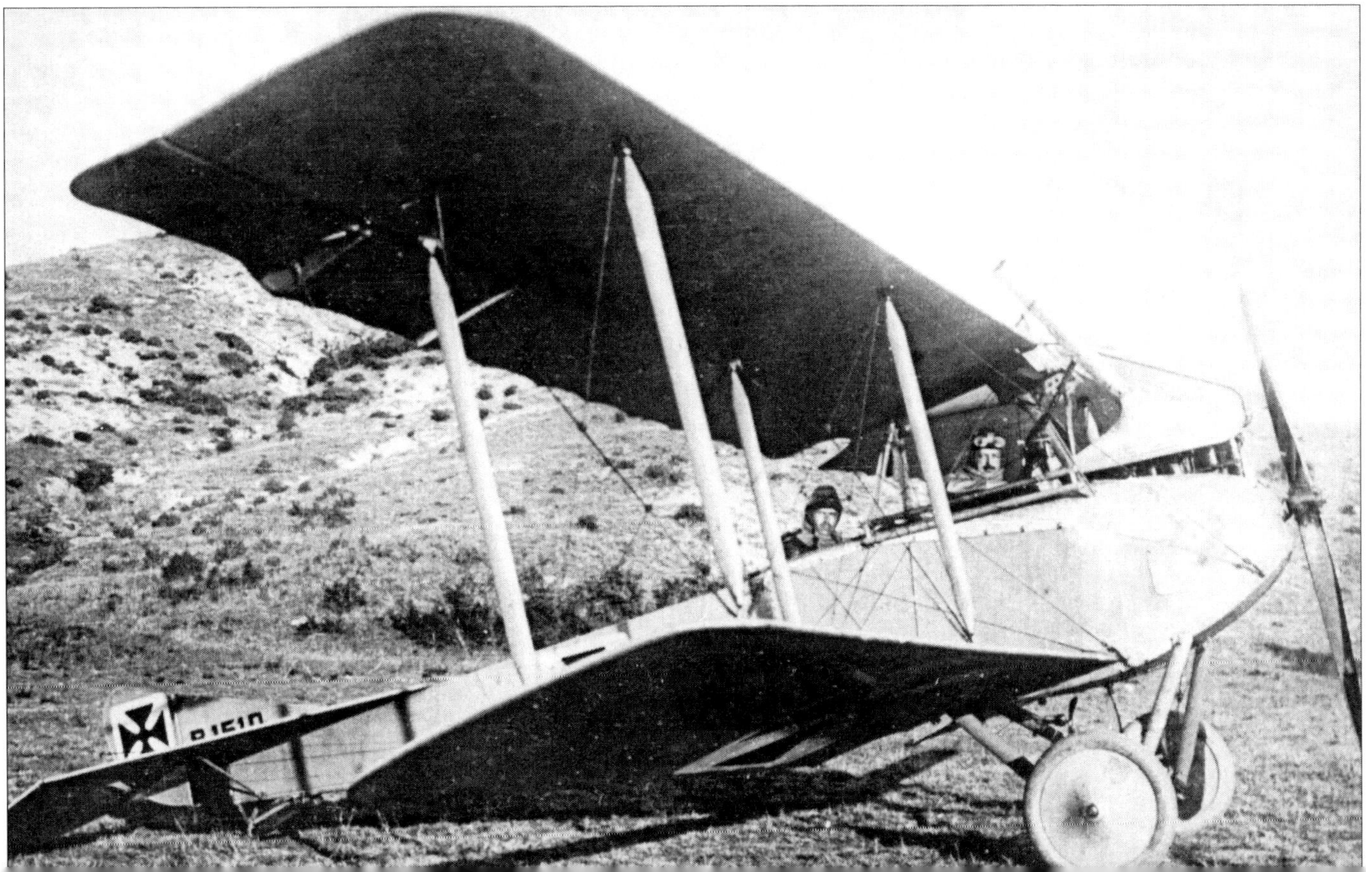

Above: LVG B.II(Ot) B.1563/15 trainer with decorated wheel covers in Germany. (Peter M. Grosz Collection/STDB)

Below: LVG B.II 534; the location and apparent lack of military serial indicate a civilian trainer. (Peter M. Grosz Collection/STDB)

LVG B.III

Above & Below: This LVG B.III is probably the prototype. The photo below shows the aircraft before the designation was applied but with national markings. Franz Schneider had left LVG before the B.III was designed so the ailerons lack the characteristic 'kink' of Schneider designs. However, like many WWI designs they feature prominent washout. The B.III trainer resembles the LVG C.V designed about the same time and has the more advanced under-wing radiator instead of the side radiators used by earlier B-types. The photo above is from the Peter M. Grosz Collection/STDB

Above: LVG B.III 3250/18 trainer. A key recognition feature of the B.III and related trainers was the over-size rudder. The wooden wheels indicate the training role and reflect Germany's rubber shortage. (Peter M. Grosz Collection/STDB)

LVG Torpedo Bomber Prototype

Above: The LVG torpedo bomber prototype built in 1915 with *Flugmeister* Herz in the cockpit. The object in the torpedo carrying rack is a test shape, not an actual torpedo. The official designation for the LVG torpedo bomber prototype, if it actually had one, is uncertain. It has been called the 'LVG 1915' and has serial C.110 on the rudder, but its official designation is unknown. With only one crewman it was not a B-type but is included here because it resembles one. (Peter M. Grosz Collection/STDB)

Above: This view of the LVG torpedo bomber prototype accentuates its distinctive appearance and highlights its size. Derived from the LVG C.I, it was powered by a 150 hp Benz Bz.III engine cooled by side radiators like the C.I. It was unarmed, had 3-bays of bracing, and ailerons were added to the lower wings connected to the upper ailerons by an actuating rod for adequate roll control with the long wings. The observer was eliminated to save weight, so it was a single-seater. The test shape is mounted and its serial number, C.110, is plainly visible on the rudder under the national insignia. Test pilot *Flugmeister* Herz is in the cockpit. (Peter M. Grosz Collection/STDB)

Right: This view of the LVG torpedo bomber prototype during evaluation shows the aircraft's large wingspan, needed to carry the weight of a torpedo. Both upper and lower ailerons have the characteristic LVG 'kink' in their ailerons. Skids were later added under the wingtips. (Peter M. Grosz Collection/STDB)

Right: This close-up of the LVG torpedo bomber prototype shows details of the complex and robust split under-carriage needed to carry a torpedo at the center of gravity. A test shape is mounted to the torpedo rack; it is likely weighted to approximate the weight of a torpedo for ground handling and flight evaluation. (Peter M. Grosz Collection/STDB)

NFW B.I

Above: An NFW B.I at Johannisthal. Designed as a two-seat trainer for use at the NFW flight school at Johannisthal, it was powered by a 160 hp Mercedes D.III and had 2-bay wing bracing. Note the decorated wheel covers. (Peter M. Grosz collection, STDB)

Below: NFW B.I N.F.W.6 of the NFW flight school at Johannisthal. (Peter M. Grosz collection, STDB)

Above: NFW B.I at the NFW flight school. This B.I has the original side radiators. The engine cowling has a distinctive 'turned' appearance. (Peter M. Grosz collection, STDB)

Below: NFW B.I from the NFW flight school. A flat radiator has been mounted over the engine in front of the cockpits. Transportation is being supplied by a horse. (Peter M. Grosz collection, STDB)

Otto B-Type Pusher

Above: Early Otto pusher with triple rudders in flight. Pusher designs were rare in Germany. This pre-war type was used for training.

Above: The triple-rudder Otto was replaced by a single-rudder design. This version had a built-in nose wheel.

Below: Otto 200 series pusher operational with *FFA* 4b with single rudder using a wooden plank runway. A few of the later Otto pushers were used during the early months of the war, mostly by Bavarian units.

Above: Pfalz-built Otto pusher with single rudder of modified shape. Pfalz-built Otto pushers only had this distinctive rudder.

Otto B.I

Above: The Otto-built LVG B.I(Ot) was an LVG B.I license-built by Otto; it was initially known as the Otto B.I until *Idflieg* rationalized the designations of license-built aircraft in August 1915. The aircraft appears to carry the black and white fuselage band markings of aircraft in units of "*Armee-Abteilung* Gaede". (Peter M. Grosz Collection/STDB)

Otto Tractor Biplane

This Page: The Otto Tractor Biplane was an original Otto design with side radiators and apparently a 100 hp Mercedes. The aircraft above has a 2-bay wing and triangular struts to brace it. The aircraft at right and below was a development of the above aircraft; it has regular 3-bay bracing that appears quite heavy and strong, probably to address weaknesses noted in the design above. Neither version had a fixed vertical fin. (below Peter M. Grosz Collection/STDB)

Rumpler B.I

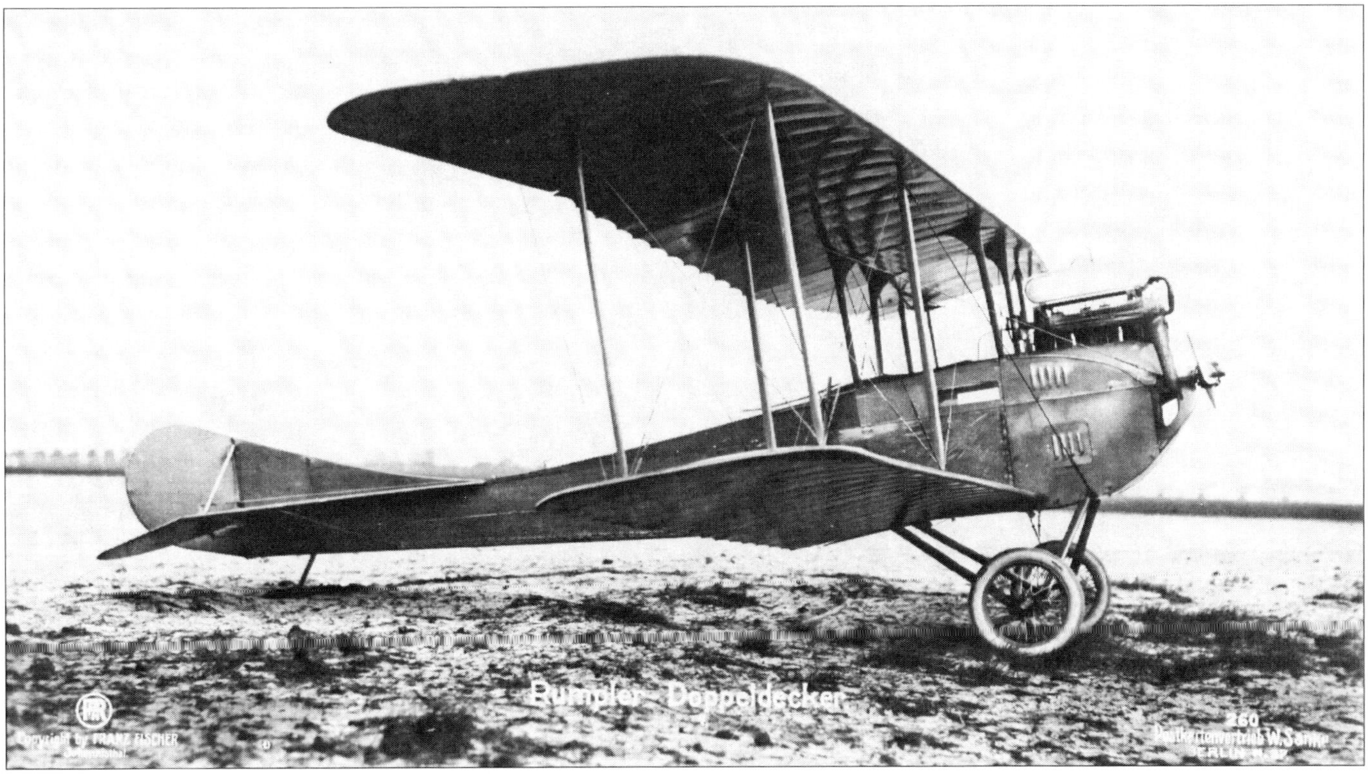

Above: This aircraft may be the Rumpler B.I prototype. It was a conventional 2-bay biplane.

Above: The Rumpler B.I was of conventional wood, wire, and fabric construction and used a 100 hp Mercedes D.I engine.

Left: Rumpler B.I depicted in Sanke card 271. Aircraft recognition, always a problem for the infantry, was in its infancy and the aircraft is well decorated with German national insignia to identify it for the ground troops.

Right: Rumpler 4A15 was developed from the Rumpler B.I by replacing its 100 hp Mercedes engine with a 150 hp Benz Bz.III.

Below: Rumpler B.I B.362/15.

Sablatnig SF6 / B.I

Above & Right: The SF6 was a landplane derivative of the SF5 floatplane intended for training. Standard late-war naval camouflage appears to have been applied. Its engine was a 150 hp Benz Bz.III.

Below: The SF6 was also known as the Sablatnig B.I. Only one was built.

SSW B.I

Above & Below: Two views of the SSW B.I unarmed reconnaissance biplane. It was a neat, 2-bay design for two crewmen. Engine was a 110 hp Sh.I counter-rotary. The B.I achieved 95 mph, a good speed for the time, and was sent to *BAO (Brieftauben Abteilung Ostende* – Carrier Pigeon Section Ostend), a code name for one of Germany's original bombing units, at the request of its commander, *Freiherr* von Thuna. The only B.I built, it crashed in November 1915 and was not repaired.

Above & Below: These two views of the SSW B.I reconnaissance biplane reveal that the fuselage and tail were closely related to those of the SSW E.I monoplane fighter. The only B.I built was sent to *BAO* at the request of its commander, where it crashed in November 1915.

Left: This view of the SSW B.I emphasizes its 9-cylinder 110 hp Sh.I counter-rotary.

Right & Below: Two views of the SSW B.I. The 110 hp Sh.I counter-rotary required a front bearing support. The Siemens-Schuckert company logo is on the hangar doors below.

Union Arrow Biplane

Above: Union Arrow biplane with water-cooled engine with factory staff. (Peter M. Grosz collection, STDB)

Below: Union Arrow biplane with rotary engine with factory staff. (Peter M. Grosz collection, STDB)

Above: Union Arrow biplane with water-cooled engine in flight. The Union Arrows were powered by 100 hp inline engines from Argus, Daimler, and NAG. The photo was made into a Sanke card. (Peter M. Grosz collection, STDB)

Below: Union Arrow biplane with water-cooled engine and Union *Taube*. (Peter M. Grosz collection, STDB)

Union Racing Biplane

Above: Union Racing Biplane. The engine appears to be a 100 hp Mercedes D.I. (Peter M. Grosz collection, STDB)

Below: Union Racing Biplane. The spacious rear cockpit could accommodate several passengers. A side radiator was fitted on the starboard side but not the port side. (Peter M. Grosz collection, STDB)

Above: Union Racing Biplane designated "U.2". There were 2½-bays of wing bracing and the lower wings are curved upward at the tips. (Peter M. Grosz collection, STDB)

Below & Facing Page, Bottom: Union Racing Biplane. The spacious rear cockpit could accommodate several passengers. A side radiator was fitted. (Peter M. Grosz collection, STDB)

Above: Union Racing Biplane with simplified wing bracing. (Peter M. Grosz collection, STDB)

Above: Union Racing Biplane "U.2". (Peter M. Grosz collection, STDB)

Above: Unknown trainer with 4-cylinder engine, bulbous nose, and tricycle landing gear. Was this an early AEG trainer?

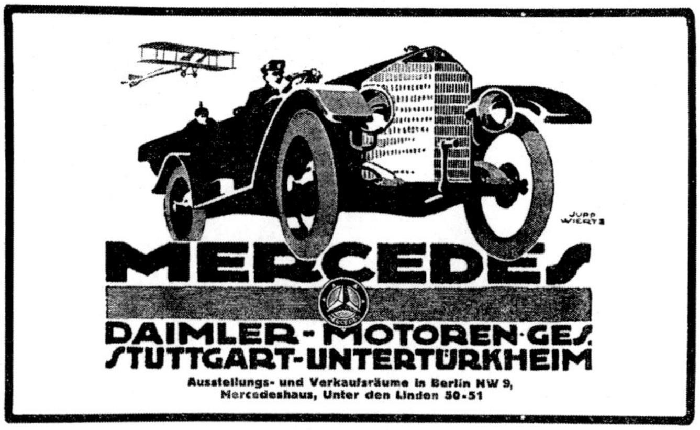

Made in United States
North Haven, CT
05 November 2026

21769185R00048